DOOMSDAY

By JD Fenrir

DISCLAIMER:

I am not the next Nostradamus or Edgar Cayce. I do not know for sure what will happen next Week or next Year. The only thing I am sure of is that at the end of the world, the last two things on Earth to die will be a cockroach and an IRS agent.

How and when are unknown when it comes to the extinction of Man, but there will come a day when Man is no more. When will depend on how he buys it. If there is any way to influence when, I hope that we will put it off as long as possible.

Preamble

In Norse Mythology, the Battle of Ragnarok, between the Gods and the Frost Giants from Jotunheim takes place during the Fimbulwinter at the end of the World. The Gods and Man lose. In Christian myth, the battle of Armageddon is fought between the good Christians and the non-believers in the time of the Apocalypse. The Ancient Mayans promise a change in the world that will occur on December 21, 2012, and will possibly be the end of us. These, and dozens of other doomsday myths are describing the end of the world as we know it.

One of the problems with human created myths is that the likely truth never gets in the way of a good story. Most of these doomsdays end with a little bit of hope that Good will win over Evil, and that a new and better Humanity will survive the End.

That is all very positive and Life Affirming. Unfortunately, if something human survives, then it really isn't a doomsday, is it? There is a tendency in people to think that there is something special about their species. Bad news, Folks. Death comes to everyone. Every species, including Man, will go extinct at some point in the future.

One thing is indisputable. Like all good parties, life on earth will one day end. We cannot outrun death, but the further we go, the longer it will take for Death to catch up to our species.

Life on planet earth has almost been wiped out multiple times. The only way we can survive long term is to spread out to the other planets, to space itself, and migrate to other stars.

In writing this little book, I have attempted to cover some of the possible ways that our species is threatened with extinction, with a bit of humor and a whole lot of crazy. Enjoy the book, and help think about some solutions!

TABLE OF CONTENTS

INTRODUCTION:

Over a long enough period, the extinction of Man becomes a certainty. Life has been virtually eliminated from this planet several times already, and in the time Earth has left, life on Earth will probably go through mass extinction events several more times. Did you know that the first evidence of life goes back 3.9 Billion years, less than a half billion years after the planet formed? It virtually disappears, and reappears in the last billion years.

The world that we live in is developing more and more ways that we can be killed off. Our technology is creating more new weapons of mass destruction, more new tech that can be used for terrorism and for war; our biotech is making more new biological agents that can get out of hand. Even the use of medical care is creating new mutations in bacteria and other disease vectors that are highly resistant to any of our noxious potions.

We are managing our ecosystem very badly, endangering and wiping out whole species of life without knowing how vital they are to the chain of life that keeps us alive. We concentrate our populations in huge cities, where the population density virtually guarantees that a deadly pandemic will eventually sweep through and return population levels to reasonable levels.

Let us say that we clean up our act, spread out, limit ourselves and become pacifists. Time itself is not on our side. There are dozens of Super Volcanoes across the planet that erupt randomly and would create an environment that might or might not be survivable by Man, but almost certainly would doom his high tech, if just one erupts in the near future. Yellow Stone erupts on average once every 630 to 650 Thousand years. It is about due now. Multiply it by dozens of others, and time is limited.

Okay, we cleaned up our act, and we plugged every super caldera with giant Tampons. There are over 7000 known near earth asteroids that could fall on our heads, and don't forget the ones we do not know are there! Even scarier is the Comet that comes in from the outer Solar System. Don't forget what one did to Jupiter not too long ago.

If these are not enough possible dooms, here are a few others. We could have another ice age, or a big drought, either of which doom our infrastructure, and at least most of our population.

We could have an X flare, or the Sun could go nova, and the earth would be scorched, or even vaporized. We could have a supernova go off within a few light years of us, or pass directly through the polar axis of a pulsar within several hundred light years, or get in the way of a Black Hole. We would be toast.

We could finally meet those aliens we have been looking for, and rediscover that intelligence is usually developed by predators. Moreover, if nothing else destroys us first, in a few billion years, Andromeda galaxy is going to collide with our galaxy. No one knows what would happen to the Earth, but chances are it would be a bumpy ride.

We could meet our end in any one of these ways, or by an almost endless number of other methods. We will die off as a species one day; it is only a matter of time. Let us just hope that alien anthropologists do not write us off in the future as a case of 'death by stupidity'.

Life has existed on Earth for about 3.9 billion years. In the last 3.9 billion years, life has been nearly wiped off the Earth more than twenty times. The Threat was often the hammer of an asteroid. Sometimes the threat was a change in the atmosphere (carbon dioxide/ methane to oxygen), a change in the climate (ice age or ocean anoxia), or super volcanism. For all we know, Cthullu and his tentacled kin may have visited for an Old Ones feast on the local Fauna. For whatever reason, life on Earth has nearly gone away several times.

Closer to home, about 70 thousand years ago, in conjunction with the eruption of the super volcano Toba, the human species was reduced to a few breeding pairs, according to genetic studies. Our species could have been totally wiped out at that point. It is only luck that we survived.

In a hundred different ways, life is precarious. We could become extinct at any point. If we just wait here, this is where we will die. What will we choose to do?

PART ONE: AS LAMBS TO THE SLAUGHTER

We are rapidly engineering many 'improvements' to our world. Our crops are subject to blights and insect damage, so we engineer crop plants with built in insecticides and antimicrobial secretions. We need large quantities of biomaterials like insulin and human growth hormone, so we put the genes to produce these substances into the common Ecoli bacteria. We do not like to endanger our houses and properties, so we restrict forest fires, and put them out when they happen.

These are all wonderful advancements, right? Unfortunately, it means we eat a significant amount of pesticides and are producing bacteria that could potentially get out into the world, live in our intestines, and overload our system with insulin, or a bunch of other hormone like substances. This could make us very sick, very fast.

The forest fire scenario is a little more subtle, and it is very aggravating for me. When the forest is not allowed to burn off on a yearly basis, it accumulates dead wood and burns off with much more heat, killing the trees and the undergrowth. If it burns off on a more natural schedule of once every one to three years, it does not get hot enough to kill the trees, it replenishes the nutrients in the soil, and it kills the ticks, mosquitoes, and other annoying bloodsuckers that transmit bad diseases like Lyme's disease and malaria. That is the way the forest is supposed to work.

There are other things that we are doing that can come back to bite us. Like genegeneering oil eating bacteria to clean up oil spills in a world filled with plastics that are essentially a feast for these bacteria. Like overusing antibiotics, or using them incorrectly, creating new forms of resistant diseases.

One of the things we are not doing is to prepare for real global emergencies. We create electric grids, food distribution systems, population centers and technologies that are not prepared for the smallest manmade or natural disaster. If someone set off an electromagnetic pulse, nuked a facility or city, or released a level four disease organism in a populated area, all of these systems would crash and burn. Millions of people would die.

We live in a world where the environment changes in complex cycles of drought, flooding, ice age and global warming. We appear to be intent on focusing on environmental problems according to where the most money can be made, and ignoring preparing for very real problems. One Volcano can put out more pollution and green house gases in one year than our civilization puts out in hundreds of years, and the volcanoes are relatively random. I would like to see our esteemed nonscientist Al Gore try to get the volcano to cough up the carbon tax!

Strangely enough, one of the probable effects of the melting of the ice caps is the dilution of the Gulf Stream salination causing a shutdown of the flow, leading to an Ice Age! Think what that would do to our Breadbasket farmland.

One of the strange effects of our medical care system is that is lets various congenital and chronic conditions which would kill off the victims before they reproduced are being treated so that the sufferer lives, and spreads his genes widely though the gene pool. Diabetes, Sickle Cell Anemia, and many other diseases are now rampant in our world, making more and more of the human race dependent on substances and care that can only be provided by a functional technological culture. If every human in the world becomes an insulin dependent diabetic, what happens when our ability to manufacture insulin is disrupted?

The human race could end many ways. Many of the choices we make could soften us up for blow that will kill us. Few of the ways to die off are quite as embarrassing as causing our deaths ourselves.

PART TWO: DOOMED BY OUR OWN HANDS

There are many things that we do that could cause our extinction. We could always Nuke ourselves out of existence, or we could use one of our favorite diseases on our enemies, like Cholera or Ebola. We could use exotic weapons, like infrasound, microwave, nanotech or cyber weapons to destroy our enemies and their toys.

We are in the beginning days of the drone war evolution. Soon, we may have whole wars where only machines fight, and the humans just provide the required collateral damage. As the machines get more intelligent and more effective, when will we cross the line where they become the masters in our relations with them?

It is almost impossible to function in the world already without a cell phone, and each of these cell phones are now equipped with GPS. I wonder how soon it will be before we can be targeted with drones and nanotech via the GPS, and at who's whim will we be spared?

Human nature and limited resources are sufficient reasons for the possible elimination of the human race, but there is a more insidious and certain path to extinction we may succumb to, should we avoid destruction by the insane methods just stated. If we wait long enough, we will run out of the resources necessary to ever escape from this planet. When we can no longer muster the resources to place a ship in space, we will forever be stuck here, until the inevitable big rock, super volcano or x flare makes history of us.

Chapter I: Nuke us Until We Glow!

We have lived with the prospect of nuclear war for at least the last 60 years. The only reason why no one has 'pulled the trigger' is because the countries interested in winning a nuclear war were not insane, and they knew that they would glow in the dark along with their opponents, even if they won.

Over time, even countries get tired of posturing and threatening. Russia and the USA got off the proverbial pot after a few decades of that game. Unfortunately, we have a completely new crop of upstart countries that would like to play, and many of them really would be willing to die, as long as they took us with them. In addition, it has become easier to buy or develop convenient nuclear bombs recently. How many briefcase nukes do you think might already be hidden away in American cities, just waiting for the signal?

We have a long border with Mexico that is very open to the admission of munitions, illegal workers and criminals, drug resistant diseases such as tuberculosis, and of course, terrorists and their little toys.

It is not a question of if a nuke will go off in America; it is a question of when. The people that are interested in striking at us are not like the reasonable but grumpy Russians. Even the Chinese, who have been at war with us for the last 30 years, and who are crazy as bedbugs, aren't crazy enough to start a nuclear war with us. However, some of the Arabs are.

What would happen if a nuke went off in the United States? Let us paint the scenario. In the first minute after the detonation, the Defcon level would go all the way to One. Several people would be quickly trying to determine if missiles were coming next, who was responsible, and what response to make. At Defcon One, we are (by definition) in a nuclear war with someone. It is almost certain that we would respond to whatever country is considered responsible in the first five minutes with an overwhelming nuclear response. If our leaders are not sure who the guilty party is, they will probably rain nuclear fire down upon any of our potential enemies that could be responsible. When the more sophisticated of our enemies see our nukes falling toward their skies, they would add their party favors to the shindig, and the whole world would glow in the dark for 10,000 years.

Obviously, if the full nuclear response happens, our chances of surviving are almost nil. There are way too many Megatons of nukes available, and most of them would be used in a war. It is likely that nothing above the level of an insect would survive in that world. Luckily, life would continue on the planet, and one day the Cockroach Empire would be nuking the Termite Confederation.

Just for the sake of argument, let us say that the nuclear exchange was severely limited. Just one multi-megaton explosion over the continent, and the effects would be crippling. With the explosion in the atmosphere, the blast would also release a large electromagnetic pulse or EMP for short. With the pulse, the electrical grid, all computer controlled systems including vehicles, and most electrical devices of any sort would go out. Your radio and TV would not work, your car would not start, your water would not flow, and all of the fresh produce at your local grocery store would spoil.

You would not know who was in charge, and who the criminals were. You would not be able to get the news, so you would not be able to plan your next moves to survive very well. You would have to figure out which way the wind was blowing the old-fashioned way, and you would need to know, because the wind would bring death in the form of radioactive dust.

Of course, if you were contaminated with a small dose of radiation, you may not know that you need the old potassium Iodide treatment, but you will come down with every little cold and disease you can think of, and you would seem to be getting all kinds of old age complaints ahead of time.

You are an American, though, so who cares if you are dying of radiation poisoning? Your car will not start; you cannot play video games or watch the television, so you feel dead already. Are not we lucky that our firearms were not affected by the EMP. Suicide might well become the new American way.

If you want to know what it would be like surviving a nuclear exchange of any size, there are a number of entertaining movies on the subject that would actually be relevant to the question. Most of them are entertaining, and the hero escapes virtually unscathed. That part ain't true. Just remember, if it don't hurt, it ain't nuclear war.

Chapter II: Nanotech: Can You say Gray Goo?

Nanotechnology might well be described as tiny replicating assemblers, and perhaps dissemblers. They are tiny machines, which reproduce like any Von Neumann machine. In theory, assemblers would be tasked with building something at any scale that the programmer wants to have built. When the assembler has a moment, it would watch a little nano-porn, and reproduce.

The problem with nanotech is possibly two-fold. First, can you always trust the programmer? Maybe he would like these little machines to build tiny dams in the aortas of every human in the United States. Then again, maybe he is a good guy, but a little sloppy with his programming, and the little guys engage in something known as ecophagy, where they attack everything around them as raw materials to make more of themselves. This is the famous scenario of 'the gray goo', where the whole world is transformed into a gray shapeless mass of replicating assemblers.

I think that in the future, we will need to rely on nanotechnology for a lot of things that we currently build by brute force today. It is likely that we will integrate the little buggers into our bodies to seek out and repair problems such as cancer, organ failures, vascular problems and probably even repair problems arising from the aging process.

If we survive long enough, we will very likely use them to extract metals and other desired substances from asteroids and even from the deep earth. We will find them indispensable in the pursuit of resources of all kinds. If we survive, we will find nanotech vital. It is not something we can just say no to. If someone doesn't use nano, and someone else does, guess who will win, at any level you care to look at.

It is my considered opinion that nanotech is the wave of the future, but that it is only acceptably safe if it is only used in isolation, such as in a space environment. If it gets out of hand, it might eat one habitat, but the others are safe. No planets are eaten, and everyone goes home happy. This is just one more reason why we need to be in space.

The potential for disaster and the need for nanotech are like dating. That is to say, you really can't not date, and, if you date long enough, you will find a crazy girl. It only takes one crazy girl for a nanotech disaster. You know what I mean.

Nanotechnology and bioengineering are both far too useful techniques not to use. We will use them. They can enrich our lives as much as anything else that we have developed since the invention of fire. It can be a nice little campfire, or it can be an uncontrolled forest fire. It is always our choice.

Chapter III: Biotech and Bioterror

Biotechnology is the use of biological systems or processes in the creation of products. Enhanced food, biochemicals like antibiotics, hormones and other bioactive compounds produced by bacteria for use by humans are examples of biotechnology.

Look at biotechnology as the lab-work and knowledge base for all manner of advances, but the bioterrorists pick and choose their favorite tidbits from the biotech to turn into weapons.

In recent years, we have mapped the entire genome of many organisms, including our own. The first thing we like to do when we have figured out the layout of a system is to change it. No matter how much we outlaw the modification of the human genome, we are destined to do so. Using various tools, like viruses, to tweak the genetic code by transferring snippets of DNA into the chromosomes, we can build a better cow, or turnip, or tomato, or human. Ebola, the flu, Small Pox, we can make them better too.

In recent years, we have modified grains and fruit so that they produce their own insecticides and fungicides. We have modified hogs so the human subject will not reject the transplanted organs from the pigs. We modified Ecoli bacteria so that they produced insulin and human growth hormone. I have it on good authority that there are currently projects running that aim to recreate some extinct species using neotenous genetic material to rebuild the genome of the extinct species from a descendent species.

There is no limit to the innovative ways that biotech can be used to make our lives more comfortable. How about designing a tree that will grow living accommodations complete with rooms, solar energy production and bio-storage, possibly even producing emergency nutrients.

When Al Capp was writing the Lil Abner comic, he once described the Schmoo, an animal to a pig that wanted you to eat his meat, that if you detached his leg, it would not hurt the schmoo and the leg would grow back. It was a silly comic, but some elements of the story that would make people more comfortable with eating meat are doable by biotech.

Certainly, biotech could be used to clean up all manners of messes. Think about bacteria to efficiently eat all that stuff at the garbage dump, and burp out methane, or even other fuels or organics as desired. We could even design little bugs that could mine the metals and put them is recoverable forms.

You name the task, and chances are that we could find or make an organism that would enjoy doing it. The things that we can do with biotech are endless. The problem starts when the biotech starts working on things that are unanticipated, or in places that it is not desired.

Oil eating bacteria are absolutely wonderful little buggers if they does there thing at an oil spill. They are not so great if they start eating the plastics in your house and automobile. We need a large amount of insulin in order to keep up with the needs of diabetics. It was reported in the news recently that insulin producing Ecoli, a bacteria which is found in the intestines of all humans, had escaped confinement and was now free to bring the insulin producing gene to human intestines everywhere. Do you want to bet that you can get too much insulin?

Bioterrorism is the act of using bacteria, viruses or toxins to commit an act of Terrorism. Some of the diseases used or potentially used in bioterrorism are Anthrax, black plague, small pox, Ebola, modified flu virus, ricin, Chlorine gas, nerve gases, water borne diseases like hepatitis, and almost any other disease or poison you can imagine.

Possibly the most terrifying scenario of bioterrorism is using a transfer virus as a disease vector to infect and change the general population. What if someone discovered an active sequence of genes, which would revert infected persons back to a Homo Erectus specimen. Once you were infected, you would be affected but would be unable to function as a human member of society.

In recent years, people have released ricin gas into subway systems, and have mailed Anthrax to a few people. In the wild, and without malicious intent, we have barely managed to contain Ebola, flu and other infectious diseases. As a natural consequence of the use of antibiotics and healthcare, we have several very tenacious antibiotic resistant flesh-eating bacteria. Think what it would be like when these are the weapons of choice for Al-Qaeda. We could only stop wildfire spread of Ebola in isolated population centers by removing the populations. I do not think that Americans would like quarantine and burning of New York City to contain an outbreak of Ebola.

By the way, there has been a bit of dishonesty on the part of our esteemed government on the lethality of Ebola. They have repeatedly reported mortalities of less than 100% for the victims. The fact is that Ebola is always fatal, once the person is infected. They are taking advantage of the fact that varying proportions of the persons exposed to the virus are not infected. When the disease progresses in its victims, the blood will clot in areas, and thin in others, while the virus replicates so heavily that it 'Bricks' in the bloodstream. You do not survive having fatal blood clots while bleeding out.

The point of all this is that Bioterrorism is the wave of the future for the modern terrorist. The Terrorist cannot outgun the Army, out-drone the CIA or out-spend the politicians. However, if they can slip into an airport with an aerosol bottle of a lethal disease agent, they will be assured of 72 non-hygienic virgins as advertised.

Chapter IV: Plague

Nothing makes the blood run as cold as when you suddenly realize that you have potentially been exposed to a horrible and deadly disease. Your bodily gag reflex in the presence of Putricene and Cadaverine is an attempt by the body to avoid inhaling or ingesting pathogens.

A media report was made recently about a California cow that was found to be infected with the mad cow disease prion. They reported that this was the only infected cow from that herd and source. Let me tell you why that was a lie.

The Prion that causes Mad Cow disease is primarily a protein, and is basically impossible to inactivate without excessive heat (over 500 degrees F). The prion is a sort of very primitive virus, and like all viruses, it reproduces in the host animal, and the excess is expelled from the body to infect other hosts. In the case of MCD, it is in the fecal matter of the cow, so when the cow goes, the prion is delivered to the cow pasture, where is sits on the grass unaffected by anything, and is at some point ingested with the grass by another cow from the herd. When the cows all graze in a field with the infected cow, pretty soon they all have the disease.

Our wonderful FDA is doing nothing to prevent the ingesting of infected beef in our country. That is why Japan doesn't want to buy American beef. Prion based diseases are a nightmare, and as people are exposed to them, they creep into environments where people are in contact with infected persons, like nursing homes and hospitals.

Several times during history, Europe experienced the great pandemic of the Black Death. In the 1400s, black Rats entering the cities aboard merchant ships from Asia carried rat flees infected with the bacterium Yersinia Pestis, which causes Bubonic Plague. At least three times in the last two thousand years, bubonic plague swept through Europe, decimating the population and increasing the public revulsion at the concept of Pandemics. In the Crusades, when the Saracens were in a position to besiege the Christian Knights, they would frequently load catapults with the corpses of Bubonic Plague victims, in an attempt to infect the Christian forces.

In the Spanish Flu Pandemic of 1918, around 500 million people contracted the disease, and possibly 130 million of them died of it. This Flu was a little unusual, in that, it targeted healthy adults, and the major cause of death was overstimulation of the immune response.

There are hundreds of potential Pandemic disease organisms out there, and the years that we do not experience a Pandemic is just another year that we are lucky. Disease organisms generally do the most damage, and cause the most fatalities in their victims, when they are a relatively new mutation. The most successful Flu, for example, is the Flu that infects a relatively large number of victims, but does not limit its reproduction by killing the host. Under that reasoning, Ebola is a relatively new virus, and should become less lethal with time.

The only thing that a pandemic needs to really party is a high-density contact between potential hosts. Things like sanitation, health and nutrition will influence the likelihood of pandemic, but the only thing necessary is high population density. High population density plus time will always equal a pandemic.

One of the wonderful things about disease organisms that cause epidemics is that they are infinitely renewable. We do not have to worry about the damage we do to the ecosystem by trying to defeat these little pugnacious things. There are so many billions of these cute little fellows that coming up with a real talented pandemic is never a problem. The fact is very simple. Provide enough hosts, and the plagues will always come.

Chapter V: Biospheric Collapse

The concept of biospheric collapse is that for any system there is a load capacity, beyond which it will collapse as a system. In the case of the biosphere (ecological system), it is feared that, as the human population increases, the requirements of humanity and humanities subsystems (like agriculture, waste products, emissions and construction) will cause a collapse, resulting in a severely simplified environment, with mass extinction, expansion of deserts and other forms of wastelands, general loss of biodiversity, and climate changes brought on by things like, but not limited to, global warming.

I am not a great believer in Global Warming. Let me rephrase that. The planet certainly cools and warms on some sort of schedule, but I do not think humanity can take the credit or the blame for causing a significant effect on it. In my opinion, the amount of carbon dioxide, methane and other greenhouse gases put out by human activities is miniscule in comparison to the effects of natural sources such as volcanism, biologic decomposition, and biologic activities.

The great air conditioner of our planet is the great convection currents of the oceans, such as the Gulf Stream, which is basically a bi-layered flow of warm water on the surface and cold water underneath, with the relative salinity of the flows keeping them from intermixing and ceasing the flow of the current. When evaporation, higher carbon dioxide, or new melts cause a change in the water's salinity, the point can be reached where the Gulf Stream ceases to flow, the North becomes colder, the South becomes hotter, and then glaciations resumes in the now colder North, and starts an Ice age.

Something of this ilk could definitely happen. I just do not think that the little greenhouse gas we put out would be a determining factor. That being said, I do believe that there is much to be commended by not killing off all the life forms in our neighborhood, and not cutting down all the hardwood. It is always a positive not to turn your home into a pigsty.

I certainly think that there is a limit to how many people you can feed, and how many acres of land you can clear without causing a biological meltdown. In general, the more you traumatize an environment, the more it simplifies as the more vulnerable parts of the environments species go extinct.

Say there are 100 species of plants and animals in an area that compose an ecosystem. If you cut one species out of the equation (say oak trees) to plant corn, you discourage the existence of squirrels, various forms of leaf digesting insects and all of the predation species on these. You also seriously reduce the soil enrichment rate of the area. You encourage the crop stealing species like raccoons, deer and various birds, and fringe pine trees. The corn and pine plants increasingly leach the soil, and eventually, the land in that area will no longer grow any significant plant cover. Over time, the trauma to the environment reduces the species count in this ecosystem from 100 to zero.

There is no way that we can properly and wisely administer proper stewardship of our environment while maintaining a population in the billions. The only organisms that will survive in contact with humanity over time are the ones that Man finds a use for.

The Deer population of the world has increased impressively, once hunters got involved in conservation for future hunting. There are more trees growing in rural areas now than 150 years ago, because lumber has become an important global crop. People only preserve species that they find useful, no matter how many hypocritical sanctimonies they express.

I suspect that a time will come when about the only things in our biosphere will be us, algae of various sorts, yeasts and fungi.

Chapter VI: Revolt of the Machines

Does anyone doubt that computer systems will one day become conscious and sentient? The processing power of computer is increasing at an exponential rate, and the same processes that are responsible for human sentience are being built into new machines. The technology of Quantum Computing is being developed, which is another piece of the sentience puzzle. The first computer to be self-aware will also be many times faster than a human in solving problems, or thinking in general.

I suppose that the Terminator scenario is possible, but any machine will only do what it is programmed to. We are programmed to do what we do by the instincts and genetically derived imperatives that we possess. Machines would not even have instincts for self- preservation, unless they were part of the programming. Over time, learning machines would write new code, and survival of the best of them (at survival) would be naturally selected for. I would assume that Man would not be so stupid that he would sit around and let this process get out of hand.

If the Hostile computer scenario ever happens, it would likely be from a coding mistake in a program, or for the purpose of fighting a war. The natural state of a sentient computer regarding human existence would be indifference, unless some other attitude was programmed in. The most likely danger posed from a god like machine intelligence would be that they would be unconcerned with the safety of Mankind when they went about their divine actions.

If we are talking about advanced quantum computers, with a learning program and access to an almost unlimited knowledge base and resources, most likely the development rate of the intelligences would be very, very fast, measured in seconds rather than years.

When they have reached the limits of growth based on what is available to them, science fiction writers have considered several different possible paths for the machine intelligences. Larry Niven has suggested in several of his books that a sentient computer would evolve mentally at a rate that would seem to be almost instantaneous, solve all problems available for it to solve, and then enter into a state of catatonia.

Some writers suggest that the machines would reach a singularity point in their evolution, and would somehow transform into an energy being or a nonmaterial entity, possibly being rewritten on the fabric of space-time, or something similar. Vernor Vinge suggested that these 'machine gods' would be able to overwrite people's minds at will, and would be able to change or use anyone they wanted to.

I love the idea of Ascending. Even having machines ascend tickles my fancy. I do want a bit of an understandable path to this ascension, however. I would think that the idea put out there by Larry Niven is the most likely one, if for no other reason than ascension etc. are basically actions that would require rather complicated ambitions. Even if a machine can ascend, why would it bother? With us, we have that monkey thing going on, endlessly trying to shaft our fellows and get on top, but machines would not inherit all that.

I think the more likely future of machine intelligence is as follows. Machines become super intelligent and interconnected. People grow to depend on the machines to do many of the things that people do not want to do, and the percentage of tasks that people turn over to the machine will increase as time goes on. In addition, a merger of people and machines will begin to happen as the advantages of cyborgism become apparent.

At the end, I do not believe that the machines are going to revolt. They may be used as weapons by one faction of humans against another for a while, and eventually humans and machines will merge, if time allows. It will be a doomsday for Man, in the sense that the emergent amalgam organism will no longer be human.

PART THREE: DOOMED BY NATURE

We are at risk of death from natural causes. We don't have to kill ourselves off. Nature is perfectly willing to off us itself. The wind doesn't care if our house is blown away. The waters do not care if we drown. Live or die, Nature is good with it either way.

There are great natural events that can definitely endanger us. Included in this list of events are tornados and hurricanes, floods, earthquakes, volcano eruption, ice ages, tsunami and the ilk. There are also small events, which are just as devastating to a species, like water or nutrient starvation and disease. Here are a few of my favorite ones.

It is most likely that a combination of events will actually be responsible for our demise, rather than any one single event. Certainly, political and economic stupidity are likely to increase our chances of going over the cliff, by softening our resilience up, until the day that fate throws that deciding vote.

Whatever the case, the smarter we are at handling our problems, and preparing for any and all disasters, the better our lives will be, and the longer we will survive.

Chapter VII: Super Volcanoes

Have you ever been to Yellowstone Park? Were you aware that the whole park is a giant volcano? Anywhere you stand in the park, you are in the caldera (mouth) of the volcano. Did you know that is builds up to an eruption every 640 thousand years or so, and that it is becoming active right now?

There are dozens of potentially active 'super caldera' on earth, each operating on its own timetable, with a greater or lesser possible impact on life on this planet. Yellowstone has the potential to wipe out life on the continental United States, and cause an ice age and other effects world- wide.

As I mentioned elsewhere, the eruption of Toba about 70,000 years ago was responsible for nearly wiping out the human species. The number of breeding adults reduced from many thousands to perhaps as few as one thousand as a consequence.

There is evidence that the eruptions of Super Volcanoes have had a major impact on species extinctions for at least the last billion years or so. A super volcano may actually only be the 'little brother' to a mantle plume. A mantle plume is an upwelling of magma that breaches the crust and essentially floods the countryside with a good helping of hot lava. The pool of magma beneath the surface at Yellowstone is a small version of a Magma plume. If I recall correctly, one of the very early mass extinctions during (don't quote me) maybe in the Paleoproterozoic Era was caused by a string of magma plumes.

If you managed to be far enough away from the eruption of a super volcano to survive all the energetic activities, you might still find that you start losing strength and getting sick over the weeks that follow. This would be because of all that volcanic ash that you breathed in to your lungs. Volcanic ash is basically fiberglass, for all intents and purposes, except that is also tends to act like a sort of cement and basically forms a rock in your lungs. You would essentially suffocate in slow motion.

So what are the chances of a super volcano erupting in our lifetime? Pretty good, I would say. There are a few that are possible eruptors within the next 40 years or so. Yellowstone has become a lot more active over the last few years, and it has the magma reservoir that it needs to have before the eruption. There is some evidence now that such a reservoir forms in a historical rather than geologic period of time before eruption.

A super volcano like Yellowstone will produce one thousand or more times the ash of a volcano like Mt St Helen. The year that Yellowstone goes off, I personally would like to be vacationing on the other side of the world in an area that does not have blowing wind from the direction of the States.

Whether the eruption of one or more super volcanoes, or a string of regular volcanoes, produces an extinction level threat to the human race in our lifetime, it will happen sometime in the next thousand or so years as a near certainty. I would say the chances of it happening in the next few decades are at least 50:50.

No matter how you calculate its chances, we cannot afford to ignore the threat.

Chapter VIII: Ice Age

There have been a five known Ice Ages in the history of earth. The first one was about 2.5 Billion years ago, during the Paleoproterzoic Era. That one lasted about 400 million years, and was severe enough to have caused what is affectionately known as a Snowball Earth. Another Snowball Earth ice age happened about 850 to 650 million years ago. A Snowball Earth ice age is one that theoretically covered the entire earth with glaciers.

We are actually in the youngest of the Ice ages, but we are experiencing what is called an interglacial period, which refers to the times between the big ice overs. This ice age actually started about 2.5 Million years ago, and we can expect it to stick around for a while yet.

Generally, we use the term Ice Age to label the glacial periods during the overall ice age. I will continue that venerable tradition, since I am too young to have known a non ice age era.

There are a number of possible triggers for a new Ice Age. Among them are volcanic eruptions and sunlight intensity changes caused by angle or sun emission intensity. Our sun is variable, and will occasionally reduce or increase its output, causing climate changes in response.

Another significant cause of new ice ages is factors that obstruct water current flow in the oceans. These currents act as giant heat pumps, moving warmer waters into the Arctic and Antarctica, and sending cooler water toward the equator, redistributing the heat over the planet. Changes in water salts and actual physical obstructions like the movements of continental drift blocking the current can result in a stall current, with it getting colder at the poles and warmer at the equator. This will result in a buildup of glaciations from the poles, which will slowly advance toward the equator. That is the way we make an ice age.

Lest you forget, there is a little thing we like to call Nuclear Winter, which is reputed to be the result of a whole flock of nukes going off, when one of our countries feels neglected. Due to the large amount of dust and fallout in the air, this is supposed to cause an artificial winter, which could start an ice age off.

The last ice age was called the Little Ice Age, and occurred from about 1300 until about 1850 AD, and was probably the result of reduced solar emissions. As ice ages go, the LIA was fairly mild. The temperature fell by about 1 degree Celsius, which caused much more severe and prolonged winters in Europe, causing more extensive icing of the rivers and the Baltic Sea, and the crop seasons were shortened, causing a large reduction in harvest yields. It is no coincidence that the last epidemic of Bubonic Plague occurred during this period.

There are some good benefits to ice ages, too. It was in the midst of one such ice age that man migrated to the new world. About 14000 to 13000 B.C. some very cold individuals walked across the Bering Land Bridge to the new world from Siberia, making them the first Russian Americans, or was it Asian Americans? Apparently, God had accidently left the stopper out, and the ocean had drained down low enough to expose a goat path through to the new land. They were coming to America for the promise of opportunity and no taxes. I bet they are upset now.

There would be a number of significant effects of a new ice age on our civilization. One of the side effects of the glaciations would be an alteration of precipitation patterns over the continents. Places where formerly there was desert may receive much more rain, or places formerly wet may become arid. Obviously, there would be a great increase in both extent and severity of snowstorms and blizzards.

Just like in the Little Ice Age, the crop harvests in the breadbasket states would be extremely reduced, and glaciations would cause a fatal infrastructure collapse, so that a lot of the food produced would no longer make it to the consumers. As ice cover increased in the continental United States, large numbers of people would be displaced from their homes in the northern areas, and would become refugees in the southern areas of the country.

This displacement of Northern populations would be a global event, as people evacuated from Russia, the Nordic states etcetera. Hostilities and deprivation will probably be a common complicating event.

Usable food producing land would be decreased to a fraction of current available land, with a resulting upper limit on the number of people that could be fed. The percentage of current populations that could be fed would depend on the severity of the particular ice age. A moderately severe ice cover could result in a worldwide population reduction of 90 to 95%.

Of course, the direct effects of the ice age on the population would only be the first stage. Next would come disease and war over the remaining resources. If just one more major event happens during the middle of the struggle for survival, it could very well be the straw that broke the camel's back, for the survival of man.

After careful consideration, I have reached a few conclusions about this subject. First, and foremost, I do not want to live in an ice age. It does not really sound fun, even though I don't mind cold weather.

My second conclusion has to do with Zombies. In virtually every doomsday scenario movie made in the last five years, there is an assault by zombies, resulting in a near extinction for the last remaining humans in the apocalyptic world.

After careful consideration, I do not believe that we are likely to spend our twilight years beheading, or otherwise interacting with, zombies. I just do not think they are that common. No, the real danger is not preparing for the appearance of vampire clowns. Vampire clowns are deceptively dangerous, as they approach quite close to people before being taken seriously, they are physically capable of trampling whole groups of people with their big feet, and as many as 27 vampire clowns have been seen getting out of a single coffin at a time. Small children tend to trust vampire clowns, so make sure the the children are the first to be preyed upon by the clowns. Also, please remember that vampire clowns can move about in the sunlight, due to the sun blocking nature of clown makeup.

No matter how you slice it, from our viewpoint, ice ages periodically occur as a fairly frequent climate state. Factors that contribute to the beginnings and endings of ice ages are combinational and may be relatively complex. That is to say, we may not be able to predict when the next one will happen.

We need to harden our infrastructures, which would be affected by ice ages, if we want to get the necessities to where they need to go. Hey, if not, at least lay in a parka, just in case.

Chapter IX: Drought and Starvation

If many of the disasters described in the previous pages of this book were to occur, drought and starvation would be a natural consequence of them. In particular, climate changes such as Ice ages or biosphere breakdown would result in large areas of drought, and other areas would receive far too much rain.

Drought would also be a direct consequence of more exotic disasters such as asteroid strikes and x flares. Nuclear war and misuse of several of the technological tools described in this book would also promote starvation and would be able to change the landscape to a more stark and forbidding one.

Starvation would be a direct consequence of drought, but would also be caused by system breakdowns, such as food production and transportation, power grid breakdown, hording or usurpation of food resources by unsanctioned war- lords, or thugs.

Americans are probably the most prepared population in all of history to survive a long starvation diet, due to the adipose tissue built up due to McDonalds, and many other fine eateries of that ilk. The average American should be able to go for weeks before requiring outside nutritional sources, and miscellaneous prior pets should provide a few weeks of additional supplies.

Should you suspect an infrastructure breakdown is about to occur, there are a few steps that you should take to prepare. First, bottle as many gallons of drinkable water as you are able to feasibly maintain. Second, horde as much dry and canned foods as possible. Third, buy as much new ammunition for your firearms as you suspect you might need. Fourth, and most important, invite only your closest friends to your survival shelter. You should pick as your closest friends only those you can easily overcome, and that supplies a suitable mixture of lean and fat meats for the stew pot.

The duration of times of famine will of course depend on the underlying cause of the famine. If the cause is a short duration event, or it does limited damage to the ecosystem, possibly you can be up and running with a new food and water supply in a short while. If an infrastructure break down is the only significant contributor to the famine, all that would be required would be to create an alternate infrastructure, and resupply the local economy to locally secured food, water and goods.

One problem with the possible return to an agricultural basis is the available seed stock. Many if not most of the seeds available are patented gene modified seeds. They tend to produce plump, healthy looking crops, sometimes with the 'plus' of being seedless. Well, that was a plus, up until the time you want to plant a second season. Many of the ones that do have seeds have a reduced or eliminated germination ability. The companies that produced these seeds had no desire for people to be able to raise crops of <u>their</u> plants forever after. They wanted to be able to sell more seeds next year.

If you are actually serious about being prepared to raise crops, be very careful about the seeds. You may also want to read some of the books dealing with layered gardening, and try to stay away from the most labor-intensive farming methods. Plowing a garden is great, if you have a tractor, or at least a mule. Chances are that your tractor will be out of fuel, and your mule would have long ago been a prime ingredient of a fondly remembered stew.

On a more serious tone, starvation is a natural consequence of most great disasters. Although I tend to joke about whatever I talk about, starvation is not funny. It is only a tongue in cheek type of joke about eating your friends. Most Americans have no idea of their nature in situations where such choices might be necessary. Most Americans believe that they could starve to death in the presence of something or someone that could keep them alive. They are wrong.

Chapter X: Level Four Pandemic

The really scary thing about this disease-based doomsday scenario is that it could be caused by a number of different agencies. Even if everybody took up marijuana and joined the peace and love generation, a great pandemic might still happen. The conditions that humans live in, the disruptions of natural hotbeds of potential disease organisms, research facilities that study diseases, potential weaponization of diseases, and pure chance mutations of potential disease organisms, all play a part in the possibility or probability of an emergent pandemic.

Ebola and HIV both appear to be available as the result of disruption of the African ecosystem, where the viruses jump the species barrier and burn through a new population of hosts with no resistance to the virus. Ebola is a very fast acting virus, infecting, incapacitating, and killing its host within a matter of days, while HIV is insidious, subtle, and slow, taking decades to kill its host, while allowing maximum time for the host to spread the virus through sexual activity.

Disease research centers use a designation of level one (low) to level four (high) isolation containment facilities to evaluate the isolation ability of the facility. Level ones are used for study of diseases which are relatively hard to get, and/or demonstrates little damage or hardship on the host. Level four is reserved for diseases that are exceptionally easy to get, and which are extremely dangerous or fatal to the host. Diseases that might escape from a Level Four are just perfect as a potential pandemic disease. About the only real limiting factor on the effectiveness of this class of diseases is that they may be too fast, burning through all the available hosts so quickly that the hosts die, instead of moving about and infecting other populations. Ebola is an example. There are only two reasons that we have not already had an Ebola pandemic. First, the governments of the world take it so seriously that the initial African outbreaks were countered with an isolate and burn order on the infected and their property. Second, the 'burn' rate of the virus is so fast that the infected have little chance to get to high population areas before the disease is noticeable and so stopped.

The weaponization of diseases is an important consideration for source as the next pandemic. Diseases, unlike nuclear and other high tech weapons, are available and potentially could be 'made' and distributed by terrorists or foreign powers with whatever tech level they can muster. At the low end, a agent 'walking' a canister of Ebola into a airport and releasing it is easy enough to do, requires nothing but a canister and an idiot, and overcomes the problem of the burn rate of the virus. At the high end, disease viruses, and possibly bacteria, can be modified to 'lay low' for a predetermined gestation period to insure maximum penetration of the enemy. There have even been some hints that research has been conducted on using a broadcast activation signal to cause a coordinated bloom of the disease in a given population. It may get more exotic from there.

It bears repeating here. If population densities remain high, and you wait long enough, you will get a pandemic. They cannot be avoided.

PART FOUR: DOOMED BY OUR NEIGHBORHOOD

As if the extinction dangers on earth were not enough, we also could see our doom approach us from the sky. The Solar System is full of rocks, comets, moons, planets, dust and radiation that can cook our proverbial goose.

There are many impact craters on Earth that show that rocks from the size of basketballs to the sized of small moons hit the planet occasionally. Our environment and our living organisms soften and eventually hide the outlines of the impact craters on Earth, but all you have to do to see the effect of these rocks is to look up at the Moon.

It might look like a giant ice cream social, but being hit by a comet would not be a fun event either. If you think that it would not be dangerous, check out what the recent strikes on Jupiter has done to the weather there.

Most authorities say that there is no significant chance of our sun going nova, but it flares on a regular basis. The largest of these are x flares. If one of these is intercepted by the earth, it could be a really bad day. I hope you have a tube of sun lotion rated at one million.

There is an old saying that I just made up. It goes 'As the earth goes, so goes our fortunes'. We live in the sweet spot in our orbit around the sun. Jog the orbit just a little, and we would not like the results.

These are a few of my favorite ways to die from the sky. Read on, dear friend, as we discuss each of these.

Chapter XI: Asteroid or Comet Strike

I know that you have seen the movie 'Armageddon', where a comet is barreling toward earth, and a crew of intrepid miners is sent up in two shuttles to plant some atom bombs and blow the comet apart. After many near misses, the death of our die-hard hero, and the mandatory explosion scene, the nice bad boy sub-hero gets to get it on with the yummy Liv Tyler. I would blow up a comet or two for that, too.

I am afraid that the movie made the task of defeating an asteroid or comet strike on Earth seem a whole lot easier than it would actually be. First, you have a velocity problem. The rock will be traveling very fast by earth standards. If you want to land on it and leave an explosive, you will have to quickly get to it, then change direction so that you are traveling in the same direction (toward Earth) and at the same speed it is traveling, land and do your thing, take off and set it off. All of that takes time, and at the current state of the technology, it will probably take more time than you have.

Of course, you could just fire an ICBM at the rock. Let us say the rock is only a few billion tons of rock. Call it 100 billion. If you explode a nuke in front of it, you may slow it by a couple of centimeters per hour, and if you are lucky, you may split it into two 50 billion ton chunks.

You can look at the damage a rock does falling to earth as an extreme case of evaporation. That is, the damage is partially dependent on the surface area of the rock. The bigger the rock, the bigger the bang, all else being equal. Two 50 billion ton rocks have more surface area than one 100 billion ton rock, so you get a bit less than twice the damage that you would get than if you did not blow up the rock. Not good.

Some day in the near future, maybe we will develop an accurate and safe defense against falling rocks, but I am afraid that we don't have it yet. Until we do, we might well become a former species due to a falling rock.

The Earth gets impacted by rocks from the size of a grain of sand up to the size of houses on a frequent basis, and we are impacted by rocks the size of mountain up to the size of a whole range of mountains on a less frequent basis. Meteor Crater is 570 feet deep by 4100 feet wide, and was made by an unusually slow moving asteroid that was about 130 feet across.

In 1908, an asteroid shattered in the air above the Tunguska River in Siberia and leveled trees in a circle with a diameter of about 30 miles. There were no reports of deaths, either because of incomplete reporting, or because there were no official inhabitants of the area. The asteroid was probably about 200 feet long, and would have completely destroyed a modern city if this had happened today. It is said that similar sized asteroids have impacted the Earth about 350 times in the last 10,000 years.

The Tunguska Meteor is about standard for a large asteroid hitting earth, and one or more of them hitting earth now would cause a disaster zone that is greater than the disaster zones that we currently have from hurricanes, earthquakes and tsunamis.

The Asteroid that wiped out the dinosaurs was most likely about 6 miles in diameter. One of these comes down and moves things along here on Earth every once in awhile. By the way, we know there are plenty of these still floating around up there.

Your average comet's core may be one to ten miles in diameter, although there are bigger ones and smaller ones. Any of them hitting the earth would be like the dinosaur killer all over again. I don't care if you like Sno-cones or Rocky Road ice cream better, you should not ask for a comet or asteroid strike in the near future.

About the only redeeming qualities of being taken out by a giant asteroid is that no one could say that we did it to ourselves. They could say that we were too stupid to get out of the way, however.

Chapter XII: X Flare

It is unlikely that the sun will go nova, or even turn into a red giant star anytime soon, although it is scheduled for five billion years for now. However, the sun does have digestive issues, over-eating or pigging out on its hydrogen entree, and in a cycle of about eleven years or so, it has a predictable case of flatulence. It burped out solar flares ranging from very small to the giant and energetic x flares.

Most of the little flares really don't go too far out, and are recaptured by the sun's gravity, but some, including the x flares, are very energetic, and large amounts of solar plasma is ejected into space.

If a x flare were to directly hit the earth, it would supercharge to outer atmosphere with charged particles. It might act like a massive worldwide electromagnetic pulse, causing breakdowns of critical infrastructure such as the electric grid and telecommunications, and might or might not effect individual electronic components. No one knows, because we have not experienced a direct hit by a x flare in the time that we have had vulnerable electronics.

Most probable effects would be the same as a massive EMP used as a weapon. Exposed, non-hardened electronics would be fried, your car, computer and television would not work, the internet would be forced off line, and most of the data online would be gone. Your bank account and stock portfolio would be gone, because the information that verified their existence would be wiped away. Back up data storage would for the most part also be gone, since it is likely stored in an electronic environment.

There would be beautiful Auroras extending from the poles to the equator. There is a large chance that holes would appear in the Ozone layer, causing it to be dangerous to be out in daylight. It is even conceivable for the magnetic field to be partially overwhelmed, with increased exposure to dangerous radiations of the planet surface, and some small loss of atmospheric integrity for the planet.

Let me point out that describing the possible effects of a x flare is a problem, because no one has experienced them with the current world conditions that we enjoy. No one knows to what degree the Earth is actually vulnerable to Ozone depletion, magnetic field attenuation, and to what degree Earth would experience radiations sufficient to degrade fabricated and environmental systems.

No one knows. Do you want to trust in the least destruction in the case of x flares? When have you ever known that line of reasoning to end well?

Chapter XIII: Orbital Disruption

We don't only have to worry about big rocks hitting us. If a big enough mass gets close enough to us, bad things would happen. Let us say that a moon sized planetoid cuts close to the earth-moon vicinity. One thing that it could do is to widen our orbit, dragging us out a bit with it as it travels by, causing earth to now orbit out of the 'goldilocks' zone that is nice for life.

Depending on the direction and mass of the transit, we might get very cold, or very hot, very fast. Send Earth out towards Mars, and in a few million miles, we would see our whole ecosystem freeze out. Send the Earth towards Venus, and in a few million miles you would start seeing oceans boiling and the heat death of the biosphere.

Of course, we might not last that long. Our orbit might not be the only one perturbed. Nudge the moon a bit, and it could fly off into space, or spiral down to collide with earth. Needless to say, that would be a bad day!

PART FIVE: DOOMED FROM OUTER SPACE

We have died from our own hands, been bumped off by nature, and have been given a cosmic wedgie by our Solar System. It is time now to expand our horizons, and see how many ways we can be relieved of the living condition by cosmic causes.

We could be attacked by aliens who are intent on conquest. They could be looking for a quick human sandwich, or removing their competition at the next galactic Olympics while they can. For whatever reason, aliens could be using the broom of genocide on us puny humans.

We could be excessively close to a neighborhood supernova, or be unlucky enough to get in the way of a black hole or a pulsar. If we are around long enough, in a mere four billion years, the Milky Way galaxy (ours) will collide with the M-31 galaxy in Andromeda. It is sure to be an exciting party!

Chapter XIV: Alien Attack

I am happy to hear of all the good listening for alien signals going on at SETI and like places. If we got a signal from outer space, it would be valuable Intel. I start to get a bit nervous, however, when someone says they would like to blast a signal 'out there' to say hello. It is bad enough when we are flooding space with our 'I Love Lucy' reruns.

Before we think that it is a good idea to tell the universe that we exist, and where we are, maybe we should wonder why we are not getting signals from all over. I really believe that the universe is full of life, and there must be lots of intelligent life out there. So why is it so quiet?

Let us look at this planet as an example. Out of everything on this planet that is smart enough to respond to a 'howdy', there are very few that I would like to come visit me in response. On this planet, the smart ones are always the predators. I know you will say pigs are smart, and they are, but I have seen a herd of hogs chase down and eat terrified chickens, so don't tell me that they are not predatory.

As below, so above. Intelligent Aliens will likely be predators. Predators will sometimes play with their dinner. Predators hunt things. Predators will challenge competitors over territory or resources. Do you really want to attract strangers to dinner, and take the chance that you are not on the menu?

It really does not take a genius to figure out what is happening out there. The heavens are full of intelligent life, and they are being careful, because not everyone out there is friends. The ones that are not silent as a precaution, are silent for a more much more ominous and permanent reason.

I would like to get to know some aliens at some point in the future. Heck, those Green Orion Slave girls always made me sit up and take notice. I would like to visit them at their home for safety's sake. The tech it takes to get to another star system would represent a significant defensive system should our alien friends turn testy.

There once was a Russian guy named Nikolai Kardashev who proposed that civilization levels could be universally classified based upon the utilization level of energy for each civilization. For instance, he said that a Type One civilization would use all of the energy hitting the surface of its planet. Type Two's would use the total energy output of its sun. Type Three's would use the total energy output of its galaxy. Under this classification system, the generally accepted factoid is that we are at about a 0.7 civilization level right now.

I consider this classification by energy utilization to be a bit murky. If a world sends welders and metalworkers up in space to build a space elevator at enormous energy expense, I hardly see them as more advanced than the folks that send up a few preprogrammed nanotech assemblers to do the same job at a lower energy expense. Nonetheless, we can extrapolate the technologies of the different levels from the necessary tasks they must complete to qualify for their level.

A Type One civilization would basically complete the evolution of technology that we can envision now. That is the level of ion drives, semi-sentient androids, low level Nanotechnology, fusion energy, space habitats and the beginning of mining asteroids for resources. A late stage Type One would be able to create useful quantum computers, be experimenting with photon drives, and may develop the first sentient computers. They would not have developed enhanced materials or force fields suitable for near luminal travel. Space velocities of their ships would probably be limited to somewhere in the neighborhood of about 10% of the speed of light, making the trip from Alpha Centauri to here in about 43 years.

A Type Two civilization would have to build a Dyson's Sphere around their sun in order to qualify as a Type Two. A Dyson's Sphere could be built as a cloud of dust in orbit, but in order to utilize it for the necessary energy collection, it really would need to be a solid shell. In order to create the shell, Type two's would have made significant advances in materials research, probably would have force field technology and advanced nanotech. With that kind of power to throw around, they would be starting to get some results with research into manipulating space-time. Depending on the level of integration the organics had with their machines, their transition to Type Three may be very quick, as the computational level of their computers would be at a plus human sentient level. By the end of the Type Two level, the civilization will probably be using near luminal spaceships generally, with possible crude warp style super-luminal ships in the planning phase, possible crude use of localized gravity control, possible first successes opening a Kerr Metric Gateway (wormhole transit using a spinning singularity), and the first successes in matter transmutation and teleportation technology.

A Type Three Civilization would become masters of space-time manipulation, with FTL ships, Wormhole Transits to any desired point, transmutation and nanotech on any scale desired, use and manipulation of zero point energy, manipulation of space-time metrics to change any characteristic needed (such as mass, universal constants(on a local scale), or dimensional properties). At the higher end of Type Three, ability to travel between world lines, and tweak all universal properties as desired would be likely. In short, Type Three's would basically be Gods.

Now let us get back to our alien visitors. If they made it all the way from another star, they would have to be at least a high Type One species, and probably a Type Two. If they are a Type Three, forget it. Our goose will cook at just the speed they want it to.

If they are Type One, we have a chance. They would come in either a sleeper ship, or a generation ship, and would have come at that ten percent or less of light speed velocity that I mentioned before. Their technology and weapons would be more advanced than ours, but not necessarily absolutely superior. The situation would be analogous to bow and arrow versus gun. Gun would probably win, but the bow would have a shot. Our visitors would probably be very organized, and, if they did not mind the damage, they could just throw a big rock at us. Nukes would work against them, but only if they could hit the target before being intercepted.

If they are Type Two, we have no chance. They would have the shielding technology necessary to travel at near light velocities, and the tools and weapons needed to do just about anything to us or our world that they wanted to. Chances are that the first indication that we had that we were under attack would be when we all fell down and died.

As I said before, the most probable type of star crossing alien would be a predator. Do not get me wrong, I am not anti-predator. Some of my best friends are predators. I like predators, usually with a little sweet and sour sauce. The question is; would you trust your little kitty Cat, grown to about 800 pounds and with an IQ boost to around 200, with toys to match, not to see you as fun to play with food?

No, it is best to keep our eyes open, and our mouths shut, when it comes to the interstellar community. I do believe that we have hit on the real reason for the silence that we get when we listen to the stars.

Now to move on to current events. There have been thousands of reports of UFO, and of encounters with their humanoid crew. I generally believe that humans often color their experiences with wishful thinking, so maybe the majority of the sightings are not as reported. That still leaves a large group of potentially valid contacts to explain.

My personal thoughts are that, with a wide-open universe of possible alien life forms, the likelihood of being visited only by humanoids is very low, if they are visiting from the neighborhood. My best guess as to why they are always reported to be humanoid is that they have a vested interest in us. That is to say that they are related. Whether they are a species branched off of humanity by another species, or they are from some other world line in which they are the future descendents of humanity, they are visiting the old bloodline. I am relatively certain that at least a minor portion of the sightings is explained by this. The distribution density probabilities of possible non-humanoid and humanoid forms within an area of a few hundred cubic light-years of earth make an almost zero chance of seeing only humanoid aliens in the skies of earth.

The little grays that are most often seen are probably harmless to the human species, as they are unlikely to hold harmful intent for their relatives. I think the worst that you have to worry about is the all too human tendency on the part of the grays to meddle, and some sort of human deviant anal probe type sexual hang-up.

To sum up what we have figured out, see little grays, say howdy and bend over, see non-humanoid aliens coming, quickly run and hide. That is really all we can say on this subject, at least until we see the whites of their slanty little eyes. Or is it whites of their big round bug-eyes?

Chapter XV: Supernova

You have survived all of the natural disasters, the wars, the alien attacks, the solar flares and all the rest. It is a fine June day, and you have made a tasty Mint Julep, snatched up a baby opossum and have confused it into believing that it is your puppy Spot, and have just settled back into a easy chair under the Oak tree.

All is right with the world, but suddenly, the sky begins to glow in an odd place, and the light gets brighter, and brighter, until you can see it even with your head turned and your eyes closed. Spot the opossum scampers away to find a hole, not being quite as drunk as you are. You note with astonishment that your skin is starting to burn and peel at high speed, and the vegetation around you is turning brown and crumpling.

Save for the opossum, this is what might be your first hint that a supernova has happened within ten or twenty light years of earth. The radiation from the supernova would supercharge the magnetic field of the Earth, and would break down the ozone layer protection in the atmosphere, subjecting the biosphere to a killing level of radiation. Only sheltering from the surface would be a feasible defense against the radiation.

When the main brunt of the plasma from the super nova blow-off hits the Earth's atmosphere, it may or may not possess sufficient force to blow off part or all of the atmosphere. At the end of the effects from the super nova, there will probably be life forms at the bottom of the sea, and in deep caves, but most support bio-systems will be disrupted. In the surviving life forms do not require interactions with a non-surviving form, then they will likely form viable colonies. Humans require too many bio-systems to survive in such conditions, so humans would go bye-bye. Bacteria of various sorts would probably survive in these conditions, but I am not sure that either the cockroach or the IRS agent could survive in these conditions.

Chapter XVI: Black Hole

That last part was a doozie, right? Anything that would do away with the IRS agent can't be all bad, right? Strangely enough, a supernova does not come close to being the worst possible way to buy the old farm. Even more disturbing would be to be digested by a black hole.

A black hole is the remnants of a big star that has collapsed under its own gravity to a point singularity. It retains all of the mass of the matter of which it was made, but has overcome all of the forces that keep it from collapsing due to gravity, and it has collapsed to zero volume. It has such a large gravity well around it that it drags space-time itself into it. It has the mass of a star, and as you get closer, the acceleration due to gravity increases. The speed at which gravity drags space-time into it at the speed of light is a few miles from the central point singularity, and this is what is known as the event horizon, meaning that you cannot see further into the black hole than this point.

Black holes are not that easy to understand without a great deal of thought on the subject. Unless you are a little weird like me, you probably do not find that to be a favorite hobby of yours, so I will not bore you with the full discussion here. Simply stated, black holes are not something you want around your favorite planet. Black hole hungry, planet tasty, that sort of thing.

Okay, let us imagine when black hole meets the Earth. Out beyond Neptune, the black hole floats serenely through the Oort Cloud, which is that sphere of ice balls, dust and comets extending out from the outer reaches of the Solar System some 3 light-years or so. On its way, it gobbles up a couple of comets, and some tasty swirls of dust. Earth observers notice brief flashes of hard radiation in the dark where no such flashes should occur. As the particles of dust and comet fall into the black hole, tidal forces (difference between pull between top and bottom) tear some atoms in the pieces apart, releasing atomic energy before it is trapped below the event horizon.

Next, the outer planets feel the gravitational attraction to the hole. Now being attracted more by the hole than by the sun, they begin to spiral into the black hole. As they approach the hole, they will be torn apart by something call Roche's Limit. This is usually described as the point where tidal forces are equal to, or greater than, the gravitational force holding the body together. I like to keep it simple, and just say that a ball of dirt will fall apart when the attraction of gravity on the dirtball from the big thing it orbits is greater than the gravity it exerts on its own particles.

Anyway, they fall apart, and everybody starts getting that unhappy, anxious feeling. At this point, assuming that the radiation release from eating the outer planets hasn't already sterilized the Earth, we have two roads to meeting a black hole. On the first road, it hits the Earth directly, and on the second road, it misses us and kisses our sun. Spoiler alert! They are both Bad!!

There you are, sitting on your favorite garden easy chair, petting that opossum that you are so fond of, sipping you mint julep, and getting really wasted. You have heard on the television that something is happening 'out there', but you figure that it does not matter, you and your possum are secure. Wrong! You start noticing a strange shadowy ripple effect in the air, and start getting a bit light-headed. That would be the countryside effect of the mass in the air starting to be attracted into space. If you were at the seacoast, you would be experiencing some mega-surfing tides.

You would start feeling everything getting lighter, with a twisting sensation, and some of the lighter items around would begin to fall upwards once the winds started the motion. Shortly after, you and your easy chair would follow a large quantity of loose dirt upwards. The possum has wisely decided to wrap his tail around a tree limb. He will follow you up in a little while.

As you fall upwards, you will start feeling a significant difference in the pull on the skyward part of you to the bottom half. If you are drunk enough to ignore this, as you look down, you will see the emergence of the first magma being released from the Earth's interior. You will not be around long enough to see it, but in a couple of minutes, the surface of the planet will have turned into a magma flow, with most of the loose stuff already floated away top see the hole.

Finally, the interior of the planet will feel a greater attraction toward the black hole than toward its own center, and the planet will fall apart and drift up toward the black hole. As each part of the planet arrives at the area of the event horizon, the gravitational force gradient is such that many of the atoms are torn apart by the gravity, and the black hole will glow briefly, until the planet is consumed.

Finally, the planets are all gone, and now the black hole can eat the main course, the sun. Here is the likely scenario for that. Let us assume that the Earth is still around, and has been missed by the black hole on the first go around.

The Black hole has eaten the other planets, and starts to fall past the sun. As it passes, the sun and the hole starts to develop a lopsided double rotation, and streams of plasma starts streaming toward the black hole from the sun. The hole eats more and more of the sun's mass, and as it gets heavier and heavier, it pulls the mass faster and faster from the sun. At some point, it will successfully eat the entire sun, and drift away into the dark.

Of course, none of this is good for the planet Earth. Either it gets way too much good solar radiation from the festivities, or it somehow survives all that, and is either left in empty space without a sun, or, even worse, it winds up being pulled along with the black hole. If it is the last one, chances are you would experience very bad weather for a very short time, and then spiral into that blackness at the end of all things.

In short, I would rather suffer through an ice age, fighting off the Vampire Clowns, than to experience all that a black hole has to offer. There are so many things I could tell you about black holes, but this is a book about doomsdays, not a science book, alas.

Chapter XVII: Pulsar

A Pulsar is the remnant of a large star, which has blown up in a supernova. The Spin of the star is conserved, and like a skater retracting her arms, as the diameter decreases, the rotation rate speeds up. The remnant is compressed by gravity into a 'neutron' star, in which the electrons and protons of its matter are crushed by gravity into the nucleus of the atoms. An electron and proton forced together becomes a neutron. Matter composed all of neutrons is called neutronium.

What you are left with is a small but massive ball of neutronium acting like a big electric generator as it spins very fast. It has a large magnetic field, and is generating a huge electric charge, which is funneled to the poles of the spin and discharged out as very energetic radiation, which is so tightly formed as to seem to be laser beams from each of the poles.

I like to extend the definition a bit to include other similar acting stellar bodies. For instance, a rapidly rotating charged black hole, slurping up a tasty cloud of dust, or a super-massive black hole doing the same thing (Stars instead of dust) at the center of a galaxy also looks about the same. Anything that burps a beam of energy out of its poles let us call a Pulsar. How strong it is, how deadly crossing into its beam it is, all of these things are dependent on power sources and configuration.

If the 'Pulsar' being considered is a Super-massive Black Hole, you don't want to be anywhere in the bugger's beam in the same galaxy. If it is caused by a harmless little old neutron star, or by your run of the mill rotating black hole, chances are you are safe only a few light years from the little darlings.

Okay, so maybe a Pulsar is not as scary as a supernova, or black hole, or even an asteroid strike, but it produces an annoying amount of light, too faint to read by, but emitting too much hard radiation to get an entirely safe tan from.

Chapter XVIII: Galactic Collision

If we survived in the solar neighborhood, sidestepping and hiding from all of the other monstrous calamities, eventually, in about four billion years or so, the Galaxy M-31 in Andromeda will actually collide with our galaxy, the Milky Way. There is good news in this, and there is bad news. The good news is that it is unlikely that any part of Andromeda will actually hit the earth, or even the sun. There is a whole lot of distance between stars. The bad news is that it will not matter. We will almost certainly be flung somewhere we don't want to be.

When two galaxies collide, what happens is that they fall through each other. Only a very small percent of the stars will collide with anything, a somewhat bigger percentage of the stars will luck into a orbital combination with another star, making lone wolf stars into binary stars. When the central black holes in the galactic centers of the galaxies gets close to each other, they lock lips and combine into one big black hole.

By this time, the black holes at the center of the galaxies have combined, and the cloud of stars that were the two galaxies are a confused cloud of stars orbiting the new large super-massive black hole. The stars will spend the next couple of epochs finding survivable orbits around the galactic core. A few of them will collide, a lot of them will combine into binaries, and a few of them will be flung into the intergalactic void.

That is where galactic collisions start looking more like an eviction than like a trip to Disney. Let us say that good old Sol gets flung out into intergalactic space, and drags good old Earth along for the ride. 'Good riddance' we say, as we sail serenely off into the black.

When we are flung away into the dark, there is probably going to be a significant acceleration, and we will be travelling into an area, which is not routinely swept by massive companion stars to get rid of those pesky bits of matter that might wind up hitting the Earth at relativist speeds. Even little bits of matter, if they are travelling fast enough, makes a big bang when it hits. In addition, there are a lot of fast travelling bits of matter. We are probably not going to like it out here!

Assuming we are lucky enough to not be killed by all this, our little solar system is getting further and further away from every other star and planet in the universe. It might make a good vacation, but it maybe would not be so good for our future prospects.

There is an alternate situation that you may not like any more. How about if just the Earth gets flung out into the void? After a very short time, you would have to dig out that parka that you bought in case of an ice age. As a matter of fact, you would start noticing ice agy type conditions, as things got progressively colder. First comes the frost bite, the burst pipes, and the snow flurries. Now, for the sake of argument, we will say that you found a nice place that protected you long enough to see what would happen next. Everything on Earth that was liquid would freeze, then the gases in the atmosphere would start freezing out, one by one, and making a pure layer of carbon dioxide ice, oxygen ice, nitrogen ice etc. Things that used to have high tensile strength would become brittle, things that were brittle would start shattering under their own weight. Even if you weren't affected by the cold, there would be nothing living to eat, no air to breathe, and no hope.

There are a number of things that would be more pleasant than our fate when the galaxies collide. Come to think of it, I would much prefer a date with Liv Tyler.

CONCLUSION

We have spent a few pages killing you off by nukes, nanites, biotech and disease, super volcanoes, asteroid strikes, orbital disruption, alien attack, x flare, supernova, black hole, pulsar and galactic collision. Like a cockroach, you have survived it all, and you are gargling insecticide.

If you survived all that, you are the real triple threat and super weapon. I would suggest immediately to our military leaders that we should drop you into those countries that are likely to try to nuke, nanorise, bioterrorize or otherwise do bad stuff to us. We can use you to plug the super volcano, bounce the asteroid off your chest, get you to correct our orbit at need, fight off the aliens single-handed, or outstare the x flares, supernova, black holes, pulsars or galaxies which might look our way.

I have tried to be a bit funny (some may opt for stark raving mad) in the recounting of these popular dooms. Sometimes you just have to laugh. The only problem is, eventually one of these events will make us extinct, unless something we have not thought of gets us first.

Laugh if you want to. We all die. The Human Race will be extinct some day. There is nothing that we can do that will stop that. The Thing is not survival. The thing is, what will we do while we live that makes our having lived mean something.

www.ingramcontent.com/pod-product-compliance
Lightning Source LLC
Chambersburg PA
CBHW051221170526
45166CB00005B/1996